AF248841

OCTOBER SKY, 1980

CRANE HOOK, 1982

GLOVES, 1980

7

STOVES, 1979

Designed by Marilyn F. Appleby, with the assistance of A. Barclay Kraft.

Edited by Kathleen D. Valenzi, with the assistance of Jane M. Brown.

Library of Congress Catalog Card Number 88-80090

ISBN 0-943231-09-4

Printed and bound in Japan by Dai Nippon Printing Co., Ltd.

Published by Howell Press, Inc., 700 Harris Street, Suite B,

Charlottesville, Virginia 22901.

Telephone (804) 977-4006.

First printing

HOWELL PRESS

ELIZA

REMEMBERING A PITTSBURGH STEEL MILL

Mark Perrott 1998

PHOTOGRAPHY BY MARK PERROTT

INTRODUCTION BY JOHN R. LANE

HAZELWOOD UNDERPASS, 1982

10

MILL, BACKLIGHT, 1980

HOUSE, FURNACE, SKYLINE, 1979

Monumental arches or great gates used to make entering a city an event, but in modern times Pittsburgh reigns among those few urban centers offering a truly dramatic experience of arrival. Emerging from the bore of a tunnel to cross a bridge spanning the confluence of the Allegheny, Monongahela, and Ohio Rivers, one is immediately confronted with the impressive, compacted skyline of the business district. Just a decade ago, however, this handsome first view of Pittsburgh was genuinely awe inspiring with smoke and fire blowing over the Monongahela River from a multitude of smokestacks—visual evidence of the apparent health of one of the world's great industrial centers. Big business headquarters and the mills of heavy industry conspicuously presented the flexed, pumped-up muscles of the traditional foundation of our national prosperity.

Today the skyscrapers remain, but Eliza is gone, reduced to a graded site. It is a rare day on which evidence of manufacturing activity can be detected further up the river in the declining or already boarded-over factory towns extending 50 miles towards West Virginia. Ten years ago there were more than a quarter of a million people working in the metals industry in Pittsburgh and its environs. Now there are less than 50,000.

The Eliza furnaces of the Jones & Laughlin Steel Corporation had stood stretched along the banks of the Monongahela River since 1859. In 1979 the mill was abandoned, and the process of its decay accelerated as did the debate surrounding its fate. With the decision to demolish the entire set of edifices and transform the 48-acre plot into a campus for high technology firms, it was clear that one of the most photogenic and historic industrial complexes in America would soon be gone.

Eliza represents the kind of subject to which Perrott has been drawn during the development of his career as a serious photographer, one that is both visually striking and resonant with human significance. In making the personal commitment to document the demise of the last steel mill within the city limits of Pittsburgh, he has lifted that inclination to a substantially higher level of artistic ambition.

Perrott is a very determined individual, but he also respects proprieties. He sought permission from the steel company to photograph the demolition process, but his request was denied. So, in the face of official disapproval and assuming a high degree of personal risk in the dangerous circumstances of disintegrating structures, he began entering the mill precincts to photograph. He continued this practice over a period of four years.

In the detritus the mill hands left behind, he sought the ghostly evidence of past human occupation—gloves, gas masks, salt shakers, upended chairs and tables, trashed files, and graffiti. He captures how the workers personalized the spaces and equipment, transforming spots within the industrial work place into surprisingly homey sites of respite from noise, heat, and dirt. Without including the image of a single individual at work, he powerfully evokes a feeling for the generations of people who worked the mills and found their version of the American dream: a source of steady income, the possibility of home ownership and education for their children, and the combination of pride in family and loyalty to ethnic neighborhood that made Pittsburgh such a tightly bound working-class community.

That particular way of life is sorely threatened now. Perrott's photographs of abandonment underline the unhappy reality that many of the people who counted on the mill economy may never work again. Already, the special, gritty, blue-collar character that made Pittsburgh distinct is fast disappearing, as the city's singular identity is transformed into a common personality difficult to differentiate from other northeastern American urban service economies. So Perrott's Eliza project is an elegy for more than a mill: the dismantling of the J&L plant is a vivid visual metaphor for changes that have overcome Pittsburgh, an image in which the vacant land the mill once occupied may be understood to represent a city's identity lost and not recreated.

Crawling through the dark buildings, witnessing the demolition of the gigantic structures, stepping back to see through the skeleton of the mill the classic vista of Pittsburgh's Golden Triangle several miles distant down the river, Perrott has documented every phase of the demise of the J&L plant. Attracted by the stunning visual properties of the industrial landscape that, over the years, have brought a number of major figures in American and European photography to Pittsburgh, his project fits at the end of a continuum of photographic expositions of Western Pennsylvania mills. The opposite extreme is the work of commercial photographer Arthur D'Arazian, who commemorated the machinery in its new state—clean, polished, and freshly installed. In the decade of steel's heyday, documentary photographers Lewis Hine and W. Eugene Smith brought the hissing, seething, working mills to life. Today, such artists as Lee Friedlander and Bernd and Hilla Becher depict the end of an industrial era.

The Eliza project should be seen in the tradition of Smith's extraordinary Pittsburgh project. This dark, epic poem created in 1955-56 consisted of approximately 13,000 negatives (among them *Smoky City*, one of the classic American photographs, in the foreground of which appears the J&L plant) and is considered by critics to be one of the great achievements in the art form of the photographic essay. Both Smith and Perrott endeavored to create bodies of work that would tell an important, revealing story of a unique locale and show an appreciation for the lives that gave meaning to it. Like Smith, not consciously seeking to make ''beautiful'' photographs, Perrott has succeeded in capturing the strength of form of the Pittsburgh industrial landscape and the character and values of the people who populated it—aspects of a place that are now becoming history.

John R. Lane

SNOWSTREAKS, 1983

SNOW, PARKWAY EAST, 1981

17

EASTVIEW, DAWN, 1981

VENT, VIEW EAST, 1980

CRANE AND SOHO HOMES, 1981

TIGER-SPOT WALL, 1982

RAILROAD SEMAPHORE, 1981

VIEW EAST PAST ANN, 1980

23

LEE SOKOL, FORMER HOT BLASTMAN

My first day at Eliza was Halloween 1935. My father had lost his sight halfway through my senior year of high school, so I had to quit school and start working to help support my family. I was eighteen when I first saw a blast furnace.

I started out as a laborer on the furnaces, then I got a job in the pig ladle house where we made pig iron for export and in-plant consumption.

Later I came back to the blast furnaces, where I started as low man on the totem pole, ''monkey boss.'' In that job I took care of the slag in the blast furnaces. Every 2½ hours or so after the metal had been tapped, you opened a cinder notch to run off the slag, which floated on top of the molten iron. When you got most of the slag off, you closed the cinder notch and opened the iron notch for the first run of iron. There was a small skimmer located in the trough between the tapping hole and the ladle. The molten iron would pass through the skimmer to the ladle, and the remaining slag, which was lighter, would pass over it and flow into a water pit.

When the furnace was dry, you had a dry blow, then the "keeper" would swing a steam-operated mud gun around, run it into the open hole in the furnace, and put a packed-clay plug in. A new cast would then start after a couple of hours. It was about three hours between each cast, five casts in a 24-hour period. The iron from our furnace went by rail car to the open-hearth furnace on the Southside.

From monkey boss, also called "cinder snapper," I graduated to first helper and then to stove tender. There were four stacks alongside the furnace, and these were heated by furnace gas with a high-speed air blower that would heat the checker work in the stoves. After so many hours, when the stoves were hot enough, we would close them up and put air into them. The air traveled to hot-air inlets called tuyeres that fed into the furnace. There was a peep sight on the outside of the furnace, and by checking to see whether a lacy crust was forming on the tuyeres, you could tell if the furnace needed more heat. There were two tuyeres per column; the P-3 furnace, "Big Annie," had 20 tuyeres at her prime.

J&L once had six furnaces, but by the time the mill shut down, Ann was all that remained. The company had put the biggest capital expenditure into her and were preparing for the long run, but economics dictated that wasn't to be.

When the P-1 blast furnace erupted, I'm telling you, that was the closest I'm going to come to being in hell. It was the worst accident I saw at Eliza. To this day I don't know where the leak came from, but iron broke through one of the hearth jackets. Iron and water make a poor combination, and when they mixed in the tapping-hole area, the furnace exploded. Other than that, Eliza was relatively accident free during the 43 years I worked there. But toward the closing days of the plant, the machinery was disintegrating.

The equipment had reached a point of no return, and it was inevitable that Eliza would close.

There was one thing about blast furnace crews that I learned: the camaraderie among us was unlike anything I found anywhere else in the mill. There was no bigotry. We had a mixture of Slavs, Hungarians, and blacks, and we got along beautifully.

A negative side to blast furnace work was trying to adjust to the heat. Temperatures on the 4-to-12 shift in the summer climbed to 162° in the shade, so we dressed with the same philosophy of Arabs. We wore layers of clothes made up of long underwear and wool safety clothing. Anyone who tried to work in shorts wouldn't last very long. Men made sure they had ample salt in their food to replace what they had lost through perspiration. There were salt-tablet dispensers at the water fountains, so we got into the habit of taking them regularly.

I became dehydrated only once. It was a strange sensation; I stopped sweating. It was around quitting time, and I had gone to the locker room to get in the shower. When I turned on the cold water, it ran over me like I had oil all over my body—that's how hot I was. I worked daylight at the time, and it took me until 8:00 that evening before I started sweating normally.

Working at Eliza made you learn to function by sound instead of sight. So many noises existed at the plant that your ears adjusted to them. If the noises changed, you knew there was a problem. Steel mill workers developed a ''sixth sense'' in this respect. In the last years of Eliza, management issued ear plugs and other ear protection, but the damage to my ears had already been done. Audio tests determined that my hearing loss was due to nerve damage.

My career as a cartoonist for J&L began with a gentleman in charge of the mechanical department. He put my drawings on slides and presented them at the safety supervisors' meeting.

I first met the Jake and Looey characters in J&L's house organ, *Jake and Looey's Journal*. I looked through the journal one day, and it occurred to me that J&L might want to use Jake and Looey cartoons. Up until that time, the journal contained only words, no pictures. I had a couple of years of cartooning under my belt, so I came up with a group of ideas and sent them to the head of J&L's public relations department. When he saw the drawings, he was beside himself with excitement. J&L decided to use Jake and Looey as permanent fixtures of the house organ, and they were first published in cartoon form in 1977.

Later J&L asked me if I'd like to use the Jake and Looey characters in a book that would describe the entire process of a steel mill—from mining iron ore to producing a clean environment. The book opportunity came as a complete surprise to me and was quite a challenge! I worked on the book during my free time. Another gentleman wrote the text, while I created the art and lettering. Of the 60 drawings I completed, 48 were used. A week after the finished drawings were delivered to J&L, you couldn't find an original anywhere. The drawings had disappeared into the inner offices and were hanging on everyone's walls.

When I left Eliza on July 29, 1979, I was the last of the long-service employees. I don't believe anyone will achieve longevity like it again. After 43 years I have no regrets. I was fortunate to work as long as I did. With everything that I've done, from my working on the blast furnace to drawing Jake and Looey, I've more than achieved the goals I set for myself. To be able to do a book was the ultimate—especially at the end of my career. I feel like I achieved a little niche in immortality.

Lee Sokol

RAIN, 1979

NUMBERS, 1982

29

VENT AND HOUSE, 1980

SINK, LIGHTBEAMS, 1982

TALBOT "TAL" HITESHEW, FORMER SENIOR VICE PRESIDENT, MELLON BANK

Even when the steel industry started declining in Pittsburgh the local banks continued to support them. Once you were a primary lender you had to stay with people. You couldn't say, "We've decided not to be your bank anymore," and walk away. Over the years we had built great personal relationships with mill management. You didn't turn your back on friends when they were in trouble.

I ran the Pittsburgh Division at Mellon Bank from 1962 until 1977. We lent money to all major accounts in Pittsburgh, including J&L. There was not a lot of research and development in the steel industry at that time. Steel manufacturing was based on old technology, so J&L borrowed money basically for working capital or to build and rebuild plants. To spend $30 million rebuilding a furnace was nothing, and five years later it had to be done all over again. Typically the lending process involved the treasurer of J&L telling us they needed money to fund a project. We then discussed terms and payback plans. For small loans Mellon Bank financed the whole amount, but since many of the loans were large—$200-$300 million—we tended to spread the financing around.

As time passed the business of making steel got skinnier and skinnier, and competition became fierce. Union and management in the United States allowed costs to get totally out of hand. Mills in America paid employees $25 an hour while mills in other countries paid $7 to $9 an hour. Steel companies in Pittsburgh simply priced themselves out of the marketplace. The Japanese proved that time and time again, and now the Koreans are doing the same thing to the Japanese.

When I was fairly new in the business, Pittsburgh had the top three manufacturers of rolling-mill equipment in the world. I traveled to all of the steel companies in the Pittsburgh area to suggest buying equipment from these manufacturers, who were crying for business, but the mills couldn't justify the cost. They believed the return on their investment wouldn't be there. I tried to tell the mills that it wasn't a question of return on investment, but one of whether they were going to be in business in 10 or 15 years. Other companies were going to buy the modern equipment in the meantime and lower their costs.

That's exactly what happened. The Japanese bought new technology while Pittsburgh watched the decline of a major industry. It was hard to see in advance. Japan started out as a nineteenth-century country after World War II, but arrived in the twenty-first century before we did. Nothing could have stopped foreign steel progress during that period.

Talbot Hiteshew

MILL SILHOUETTE, WHITE LIGHT, 1980

34

FAN, 1983

35

RAIN, TORPEDO CAR, 1979

VIEW THROUGH CRANE WIRES, 1980

BOB "IKE" EISENGART, FORMER HOT MILL ELECTRICAL DEPARTMENT FOREMAN AND GALVANIZING DEPARTMENT MASTER MECHANIC

J&L's Pittsburgh Works were written off as far back as 1960. The company always talked about improving or rebuilding, but it never happened. Those of us working out of Pittsburgh felt that our mill was just a giant testing ground for J&L. Management would decide to try something new, like stainless steel. We developed the process and rolling practice, but the work would be given to the Aliquippa or Cleveland mills. We resented it, especially since employees at Pittsburgh were being let go because there wasn't enough work.

When I started at the hot strip mill on March 1, 1947, I was a young punk, twenty years old, fresh out of the Marine Corps. My first job was as a Utility B-man—someone who could operate cranes or work in electrical maintenance. After six months I was permanently assigned to the electrical maintenance department.

Once assigned to a crew, you operated on the "buddy-buddy" concept. We worked so closely together we were able to be tremendously efficient. When my crew first started together, it took us almost 24 hours to change the ¾-ton wheel on a crane. After a while, we worked so well together, we could change it in about 2½ hours.

The Eliza blast furnace shop made molten iron, which was poured into large railroad cars that we called submarines or torpedoes. The cars carried the iron across the river on tracks to the Southside, where the iron was poured into open-hearth furnaces and processed. From there the molten iron was poured into ingots, which were lowered into soaking pits to stabilize temperature before being sent to the blooming mill. At the blooming mill the ingots were squeezed, turned, and rolled into long slabs of various widths. These slabs were run through a shear to cut them to size and taken outside to cool before being reloaded onto railroad cars and returned to the Pittsburgh Works.

I worked at the Pittsburgh site for 35 years. During that time there were several strikes. Once the Monongahela Connecting Railroad, which had offices at the Pittsburgh Works, sympathized with striking coal miners and refused to move our steel. Our union didn't want us to cross their picket line. I was supposed to be working the midnight shift. None of the union officials could tell us what was going on with the coal miners, so I decided to give the strikers one more day. If nothing changed, I was going to cross the line regardless of what the union wanted. I had a family to feed. I was for the union, but it was ridiculous to keep us from working when they didn't have any information on the strike. Fortunately, when I went to the mill the next day, the strike was over.

I never went to union meetings, because they always turned into shouting matches and nothing would get accomplished. Whenever the union went on strike, money was always the issue. If the workers walked out, J&L couldn't meet the needs of its customers, who would take their business elsewhere. That cost J&L money. Strikes were also extremely expensive because equipment deteriorated quickly in the mills when it went unused. As a result, the steel industry and union formulated the no-strike system, but when the industry agreed to give employees a 13-week vacation, they gave the whole store away. Today, management realizes you can't do that and still have a business. I blame both the unions and mill management for being too greedy.

Over the years, a lot of fellows got hurt, so safety was very important. Our accident rate was low compared to the butcher-shop atmosphere of the early mills, where you'd hear about a guy getting his leg cut off and another man stepping up to take his place— no reports or investigations.

At J&L we were required to conduct safety meetings at least once a week. Foremen tried to see every worker during each turn to make sure they were wearing their gloves or goggles. J&L wanted us to keep safety first in our minds, so they created a big safety program to instill safety in people and make them be more careful.

But no matter how careful you were, there was danger. Sometimes we felt the safety measures actually created hazards. When the safety program was improved, we were required to wear hard hats. We in electrical maintenance hated them with a passion, because when you worked on something up in the air, the hard hat would fall off each time you bent over to yell down instructions or pull up a rope. If you tried to grab the falling hat, you risked losing your balance and falling too. If you didn't grab the hat, you risked having it hit someone standing below. What we started doing was taking our hard hats off when we were up high and putting them back on when we were finished.

Luckily, there weren't too many bad accidents. There was one fellow who was working on a turnaround—a big mushroom-shaped machine that lifted slabs of steel and pivoted them a quarter turn before a pusher moved them down the line. In the turnaround area, there was a sluice that carried iron scale into a large pipe running underneath the floor out to dross tanks in the slab yard. The sluice, about two feet deep and three feet wide, had a tremendous volume of water going through it all the time. Iron scale was heavy, so a good flow was necessary to wash it through completely. Somehow this guy slipped and fell into the sluice. By the time they pulled him from the dross tanks, he was dead.

Another fellow was unloading a lift of sheets onto a buggy after it came out of the pickler tank. He stood behind the lift while the crane raised the chains off the carrier, but the bullrings got caught on the eyes of the carrier, and the lift of sheets kicked over and crushed him.

Whenever an accident occurred, we had to write a report. The safety department would then conduct a review, walking through the mill to inspect the area. Regular safety tours came to be a joke, because what we considered to be serious hazards, the safety inspectors seldom saw, even if we pointed the hazards out to them. The inspectors always claimed it would cost too much money to fix the problems we pointed out. Instead, they would make a big deal over a red light being out on a crane.

There was a lot of horseplay in our department. In the electrical maintenance shop there were two fellows named Big Al and Charlie. They were like Mutt and Jeff. Big Al looked like a Greek god—handsome, young, and shoulders like Lil' Abner. He was an ex-Marine and in excellent physical condition. Then there was little Charlie who was a real smart guy. He had been a laborer before he came to work in our department. Once he joined our group he learned quickly and became a good electrician. Charlie and Big Al became good friends and would always play around. One time Big Al picked Charlie up and hung him on a hoist hook by the back of his belt. As soon as Big Al let go, Charlie's belt broke, and he fell to the floor. Big Al tried to grab Charlie before he hit the floor, but the hoist hook was swinging and it smacked Big Al on the forehead. It was the funniest thing to see. Luckily, neither one of them was hurt.

As time went by and things got worse, people had to move from the cold mill to the hot mill, or vice versa, just to have work. LTV Steel had purchased J&L, and they started whittling away at its employees. First to close at Eliza were the blast furnaces, and then the hot mill. I was there for the last rolling. As the coils were processed through the cold mill, I watched each succeeding unit go down. Boy, when that big, old hot mill shut down and the last red-hot bar went through, we all knew it was the end. The up-river crane man started blowing his siren, and then the center crane man started his siren, and the next crane man started his siren, until all of the sirens were blowing throughout the whole mill. It was a roaring madhouse. It was heartbreaking—big,

husky steelworkers had tears in their eyes. It was the end of an era.

I was one of the lucky ones. I stayed at the mill for a little longer in the galvanizing plant. Although it wasn't making a lot of money, it was in the black. Then one day management put out the word—galvanizing was going to shut down too. We prepared to leave, but J&L gave us 10,000 more tons of steel to roll out, so we stayed in operation a little longer. We'd get reprieves like that two or three times. Each time we thought it might keep going, but it didn't.

When the end came, we held a big tool auction, and people from remaining J&L departments bid on them. To make it simple, I told everyone, ''Pick a number out of the hat and whoever's number gets pulled first will have first choice. If you want air tools, you'll get all the air tools. We're not dealing them out one at a time.'' We auctioned off everything that way.

Then we had to start cleaning up. I had

a welder put pipes around all of the low areas and string cables up between the pipes so no one would fall in and get hurt. We did whatever was necessary to wrap things up.

I had told my guys from the start that we were going to do a good job so we could walk out on that last day like gentlemen, with our heads held high. And we did. We left that place nice—nothing flammable, no bottled gas, no oil, no lumber, no nothing. I felt good about it. Up until the last minute, we hadn't shied the company out of one minute of work.

Bob Eisengart

CAGE AND LADDER, 1979

SNOW-COVERED TANK, 1980

45

TWILIGHT, 1982

46

GAS MASK, 1981

HARRY PETERS, FORMER MACHINE SHOP FOREMAN

The main purpose of the Eliza machine shop was to repair worn and broken parts of different equipment, to fabricate and machine new spare parts for the various Eliza departments, and to have the other shop's craftsmen service a variety of needs. Our areas of responsibility included the blast furnaces, the ingot mold foundry, the power house, boiler house, pump house, as well as other departments in Eliza and in the Pittsburgh Works. In our shop we had some of the largest machine tools of any other machine shop in J&L. We also machined large bells and hoppers for the Aliquippa and Cleveland blast furnaces.

At one time we had more than 70 people working out of the machine shop. Besides the machinists and the apprentices, there were safety apparatus inspectors, millwrights, tool room attendants, spare parts expeditors, fire apparatus inspectors, receiver-shippers, mobile equipment operators and repairmen, cranemen, janitors, and a labor gang.

Our shop worked three shifts so machinists would be available 24 hours a day in case anything was needed for the blast furnace and the utilities departments. We maintained a regular schedule, except on Christmas, when we would ask for volunteers to work. Those who worked could have New Year's Day off.

One of the things I was most proud of in our shop was our apprentice-machinist programs, which over the years had trained a fine group of journeymen machinists. The selection of our apprentices was controlled by a committee that had developed a fair, competitive selection process. Those applicants with the best grades were offered jobs, but their being accepted did not come without sacrifices.

Applicants who had higher-paying jobs when they applied actually lost money by coming into the apprentice program. They took a cut in pay, had to buy books for the classroom work and machinist hand tools that were required to work their trade, and were expected to work three turns. Once their training was complete, however, the dedication and hard work paid off. Machinists were some of the highest-paid workers in the steel industry.

At one time every steel company had its own programs for training craftsmen. That's one reason why mills had great craftsmen. As employees left one company for another, ideas were passed along and new ways of doing things were learned. Today you shudder, because you know the education new craftsmen are receiving is inadequate. You can have all the computers and high technology you want, but you still need one thing—good old common sense. The only way you learn it is by intensive, on-the-job experience.

The people that worked at the Eliza machine shop were a proud, close group with a great pride in God, their family, J&L Steel Corporation, and their jobs. When you worked there for more than 30 years, like I did, you started thinking of your co-workers as family. We went to each other's christenings, weddings, and funerals, and we grew to know each other's children. When a son or, in later years, a daughter was old enough to go to work, the father would come to you and ask your help in getting them a job. This meant we often worked with two generations of families in the mill at the same time.

The Eliza machine shop was one of the first departments at the Pittsburgh Works to shut down, and J&L told us the date that it would occur months in advance. We appreciated the early notice because we all had personal matters to consider—car payments, house mortgages, as well as planning for the future.

A few days before the machine shop closed, all of the company hand tools were packed in boxes and crates to be distributed to other J&L departments and shops. The shop's own tools and machine surfaces were coated and wrapped to protect against rust.

On the morning of the last day, we drove our cars to the shop to load up our personal tools to take home. During the entire shutdown period nothing was intentionally damaged or broken; most employees felt no animosity towards J&L. I believe our cooperation and goodwill was a result of J&L's having given us plenty of time to prepare for the shutdown.

On that last day J&L let us host a big, lunch-time party. We invited the entire Pittsburgh Works. The carpenter shop supplied planks for seats so that everyone could sit down and enjoy their lunch and the program that followed. Our people brought food from home, and canteen workers supplied food and soft drinks. Venders that had supplied us with goods and services for years also provided catered food, as well as money for paper plates and plastic utensils.

The party was a huge success, but for many of us it was like a wake after a funeral. We were losing a very dear friend that had been a part of our lives for years. There were a few tears, but there were also many hugs and handshakes. You have to experience something like this to understand the conflicting emotions we all felt. My memories of the Eliza machine shop will be with me for as long as I live.

After I retired from J&L in 1979 at age 56, I went to work for a local manufacturer of steel mill equipment. I worked with engineering departments and other people in steel mills all over the United States and at many offshore steel companies. In my travels I discovered former J&L employees working in responsible jobs at other firms, but in general, I seldom found a company with the same caliber of people that once worked at J&L's Pittsburgh Works.

Harry Peters

HANGER AND GASKETS, 1982

STRIPES AND CHAIR, 1980

53

PULPIT, 1982

CIGAR BOX, CUP, CAN, 1982

55

BILL HIROSKY, FORMER FURNACE COOLING ATTENDANT
AND USWA LOCAL 1843 ZONE GRIEVANCE COMMITTEEMAN

When you walked through the demolition site where Eliza used to be, it was really haunting. Over the years ghost stories about blast furnace employees who had died tragically at the mill had developed. Some claimed that they heard their voices around the furnaces from time to time. It gave you chills. Tons of superstitions surrounded the mill. There were even some write-ups in the local papers about these stories. You walked through there—and it's not like you ever saw anything—but it was just spooky.

Generally, the relationship between workers and management was good. There was not a lot of mistrust between them, unlike other mills. It would be misleading to say we were completely open with each other, because that kind of trust is rare. Managers, especially line foremen who had direct responsibility over operations, had very good relationships with the rank and file. However, the higher the management, the less empathy you'd find for the workers.

When grievances and complaints did occur, the zone committeeman from the department handled matters. Whenever I received a grievance, I made a point to see the person who filed it and give him a quick estimate on what his chances were. If he was totally wrong, I'd point out to him that it was not a valid complaint and give the reason why. Most workers understood that.

Fights sometimes broke out among the workers—hourly and salaried. Usually the fights were about how to do the job or simple personality clashes. No one ever got fired that I know of. Fighting generally was never reported. Even if it were reported to top management, they would rarely do anything about it. What could they do? These were grown men. Workers were protective of each other, and management was protective of the workers.

If a worker came in drunk, his co-workers would just absorb his job, unless he made a habit of it. If the job wasn't getting done, the supervisors would take the worker off the job under the contract's relative-ability clause. The union usually didn't fight these cases, because if the guy held a critical job, like checking gas masks or oxygen tanks, and both the company and the union knew the guy was incompetent, the two sides, as well as co-workers, demanded that the employee be removed for safety reasons. It was a tough situation, but it came down to whether you wanted someone killed or wanted the job performed well.

I started at Eliza as a laborer in 1964. My progression up the mill ladder wasn't planned, because I was thinking about different things. People were being drafted every year, and I was trying to stay out of the draft and work and make money. But finally my turn came, and I was drafted. When I returned from military duty, I went to work for J&L on three shifts that rotated around the clock. That schedule was extremely hard to adjust to. Multiply the effects of jet lag by 10, and that's what rotating shifts felt like.

One department used to have an 8 p.m. to 4 a.m. shift, and when workers finished, they had tailgate parties up on Second Avenue. They'd have their beer in coolers and sandwiches laid out. Then around 5:30 in the morning, they'd go home. Those were Pittsburgh's first tailgate parties, long before the Steelers made them famous!

I decided to become a union officer because of one supervisor. One Sunday I was very sick and couldn't come to work. It happened to be a pretty nice, pleasant day, and a dozen or more employees did not report to work. I had missed only a couple of days in two or three years, but after I returned to my job, the supervisor harassed me for months. He thought I had played hooky to take advantage of the weather. I figured I couldn't fight him because he was kind of big, so I decided to run for a union office to get back at him. I was persistent. I took a few courses in labor management and educated myself in order to get a better grasp on what union officials did.

A lot of laborers thought that union officials were raking in the money, but they had the wrong perception. As a zone committeeman, I was making about $150 extra a month for about 150 extra hours of union work each month.

In the industrial field at that time, changes were coming hard and fast. J&L had job eliminations frequently, which was typical of the steel mill business. As machinery needs became obsolete, bodies went. Management tried to find what they called ''suitable, long-term employment'' to replace employees' old jobs, but they couldn't always find the right job. During those last years in the mill, most of the workers had anxiety problems. When faced with the choice of retiring or working a new job, most took their pensions and TRA—Trade Readjustment Act—benefits, which gave them an additional $250 a week.

J&L gave the union three-years' notice that they were going to shut Eliza down. Then the department started setting all kinds of performance records, which created a belief among workers that the mill wasn't going to close down after all.

When management began to raze the Eliza site, workers stood around to watch the demolition. One or two furnaces came down at a time, until only the newest furnace, Big Ann, remained. Once again, workers thought that the place wouldn't be shut down entirely because they'd left Big Ann standing. That furnace too was eventually destroyed, but up until the last minute, people continued to believe management would cancel the demolition. They were convinced that the government would stop the destruction in order to keep the blast furnace on hold in case of another war.

In the 10 years since the blast furnace shutdown, both union and management have learned some valuable lessons. Working together now is a necessity.

Bill Hirosky

VENETIAN BLINDS, 1980

COFFEE POT, 1980

TEAPOT CLOCK, 1982

SALT SHAKER, 1982

FOREMAN'S OFFICE, 1980

64

SOHO IN, 1982

LEROY BURRELL, FORMER SLUDGE FILTER OPERATOR

One of the first things you got when you started at the mill was a safety talk. They told us about hard hats, steel-toed shoes, and other safety measures—looking out for engines, cautionary road tracks, and gas. Carbon monoxide gas was one of the main problems around a blast furnace, and we learned to act with caution.

For instance, one day about five of us were coming down from the ladle house, and we saw a hot blastman stumble and fall. If any of us had rushed in after him, we could've been gassed. Instead, we held our breath, dragged him out, and took him to the hospital.

I worked in several different areas of the plant—the furnace, ladle house, stock house. I even drove a tractor, operated a crane, and was a safety man, but the best job I had there was sludge filter operator, or dorr thickener, which I did for close to 22 years. That job involved taking waste off the furnace, which was eventually recycled into pellets that were reused in the furnace. I went through a lot of trouble learning how to do that job. I got in people's way trying to learn how things were done and what the machinery was called. Sometimes jealousy played a big part in older workers not helping younger workers. Like some of the old-time stove keepers, they were hard to get along with. You could get along with the boss better than you could with them. Generally, though, if you stood around and watched people work, eventually you'd pick up things.

The sludge filter operator was a good job all in all. I was by myself, but I never really felt lonesome. The ladle house was across from me, and the furnace was in front of me. I knew my job, and the boss knew I knew my job. If I told my boss that I thought there might be a problem somewhere, I'd stay put until it was corrected. They even called me at home sometimes to ask me to come in and straighten out a problem. My co-workers used to tease me that I messed things up on purpose to get the overtime!

There was one thing we had to watch for in the mill—rats. They came up from the river and were so big they looked like small dogs. I remember one time I had tied my lunch up high—this was before the company gave us Frigidaires—to keep it safe while I went to do something. When I returned, I saw a rat leaping at the bag until it knocked my lunch down. Then four others ran over. I stamped my feet to try to scare the rats away, but they just turned on me and started chattering. I backed off! J&L finally had to bring a company in to get rid of them.

During lunch, my friends and I would sometimes sneak out and play cards. Once, one of the guys dropped a card on the floor and bent down to pick it up. In the meantime, a steel beam came down from the ceiling, and when he straightened back up, his head went ''Boing!'' on the beam. Everyone got quiet because we thought we were going to have to take him to the hospital. After a moment, he just shook his head and said, ''Ooo-wee!''

We all started to laugh, and I told him, ''Boy, remind me never to hit you on your head!'' It was like a brick.

Another time, I played a trick on one of my friends. I had found a string, so I threw it on him. He thought it was a snake, and we had to grab him because he was running around like crazy. I thought he was going to kill himself. I always felt a little bad about that one.

When I left the sludge filter operator job, another guy was hired to step in. I happened to be downstairs in the office when my replacement came running in, saying, ''There's a ghost up there!'' He was too scared to go back. I figured he probably heard a funny noise from downstairs, because I had worked up there for a long time myself and had never seen any ghosts.

To me, if you've got a group that can work *and* take jokes—a lot of guys talk and then they can't work—the day goes along nicely. The boss knew who worked and who didn't. You were wrong if you thought you could get one over on him. If you had a pretty decent boss, most guys wanted to show him what they could do.

There were some who goofed off. For instance, when you worked with slag, or when the iron ran slow and started to build up in the runner, you needed a little help getting the heavy stuff out. We had some guys who would run to the toilet, thinking that when they came back, the job would be done. But we'd just wait until they'd get back. If someone did his work, you didn't mind helping him out when you had a little free time.

When nothing else needed to be done, I would read. I would pick up anything and read it. Some of it was very educational. One of the nurses there liked mystery novels, and she and I used to know all the authors.

I never really had any racial problems at the mill. Back in the days when I started, either you fought and got over it, or they fired you right off the bat. If you had a disagreement, they'd tell you to go outside and settle it somewhere else. As a whole, everyone got along pretty well. I'm just like anyone else—I'm not going to show but so much of myself. I always respected a guy who could come in the mill with the attitude of wanting to advance no matter what the starting salary or position. I hated to see someone come in and demand something. A few were like that, but the majority would work any job.

I still keep in touch with guys from the mill. We talk on the phone, and I meet one of them downtown every other Monday. We have a couple beers and a sandwich and talk about what's new with the insurance plan or if someone is sick. He's the type that keeps up with everything that's going on, so he keeps me informed. I go out with another buddy of mine about once a month. Seeing those guys keeps me going—it keeps me busy.

Leroy Burrell

CONTROL PANEL, 1980

70

BLAST PIPE DETAIL, 1980

WINCH, 1980

TROUBLE ON TOP, 1982

73

NANCY ALBERT, FORMER LABORER AND LARRY CAR
OPERATOR

People are usually pretty amused by the fact that I worked in a steel mill, but in Pittsburgh it shouldn't be that big of a deal. Tons of people from this area have worked in steel mills.

I got my start at Eliza through a big recruitment drive for women at the plant while I was still in college. At that time the steel companies had been ordered by the courts to hire a certain quota of women and minorities. A bunch of women, including me, said, "Why not?" because we'd be making more than 10 or 11 bucks an hour.

My first impression of a mill office was that it looked like a military compound, just a mess. We went in on the daylight turn the first day, but we really didn't do anything. They assigned us our clothes and showed us where our lockers were. Most of us didn't know anybody at the mill when we started, which was uncharacteristic, because rampant nepotism existed at all steel mills at that time.

The first turn I ever worked was midnight.

Walking into that place for the first time at night was really imposing—sparks, noise, dirt—it was a humbling experience for a college student.

To watch them tap a heat or see the molten iron was really something. Actually, with all of the technology available, steel making is still a very primitive process.

Not all the women working there were athletic like myself, but if a woman couldn't handle the job, the guys would take care of it for her. A lot of the men were old enough to be our fathers, so they wanted to try and help us with our work. Women who were able to hold their own weren't resented.

You're always going to have a few idiots, but on the whole most of the guys couldn't have been more gentlemanly. They liked having women around. They had previously worked in an all-male environment, and we added a little bit of variety. Some of those jobs, especially production jobs, can get pretty mundane. Having women around made it a little bit better for morale.

My father, on the other hand, did not want me or my sister Rosie working in the mill. He didn't object to having women in the mill; he just objected to his daughters being there. He felt someone might give us a hard time, plus it was dangerous work. If you even walked the wrong way, you had the potential for getting hurt. Each department was big, and you really didn't know what was safe and what wasn't, so most workers didn't venture out of their assigned areas.

That summer I worked on the ore trestles, in the stock house, and ended up driving a larry car. The stock house, which contained iron ore and other raw materials used to charge the furnaces, was located underground, beneath the ore trestles. Larry cars would pick up loads from the stock house bins and carry them to the hoppers that fed the furnace. A larry car operation is constant; as long as a furnace operates, the larry cars operate. Sometimes the stock bins beneath the ore trestles would not be in the right place, or the doors would be stuck open, so that when ore was released from the bottom of the railroad cars into the bins, it spilled on the ground. In that situation I had to go underneath the trestle and shovel out mounds of iron ore. It was pretty physical work.

I graduated from college in 1977 and took a job at J&L that summer in security. I worked the gates, kept track of trucks coming in and the movement of shipments, walked the perimeter of the plant, and checked alarm systems. I had occasional drunks to get rid of, but there weren't that many problems. Most of the people who worked at the mill weren't going to do something that could be detrimental to their jobs.

Next, I worked in the security office as the superintendent's clerk. About nine months later, I moved to personnel, where I worked for a couple of years while I got my Master of Arts degree in Industrial Relations. When my boss left the Pittsburgh Works to go to Aliquippa, I took his place and became the supervisor of personnel and employment. I've been there 12 years now.

When Eliza went down in 1979, J&L pensioned anybody off who was eligible. Those who weren't eligible had rights to transfer into different departments if they wanted. The strip mill, which had about 600 to 700 people, went down in 1981. Later cold finishing went down, the bar mills went down, and finally the front end at Southside was wiped out. Generally, everybody who was eligible and who wanted to take their pensions did.

My job in personnel during that five-year period was very painful. It became more and more difficult to reassign people to other departments, and I was not able to get jobs for everyone who needed one. A lot of people got lost in the shuffle; they ran out of benefits and weren't trained in any other kind of employment. Many older workers had been immigrants and could barely read, write, or speak English.

The shutdown had a definite impact on the city, because it involved a majority of the population. However, Pittsburgh is starting to come back; we are not as dependent on steel as we used to be.

Nancy Albert

BATHROOM, 1982

78

TUNNEL, 1981

79

WAGON, 1983

PIPES, 1980

TREE AND FURNACE, 1980

LIGHT BULB, 1980

83

RAY FLEMING, FORMER PITTSBURGH WORKS CONTROLLER

inancial people really don't know every-thing that's going on. We put numbers together to try to paint a story, but the full story is out on the mill floor where the operating superintendents and foremen are sweating over their problems.

From January 1970 through September 1972, I was the controller of the Pittsburgh Works and supervised the entire financial operation of the plant. That included everything from day-to-day record keeping, financial planning and forecasting, analysis of results, presenting information at cost meetings, and keeping the works manager and superintendents informed. I had been assistant controller there for four years prior to that.

Our accounting personnel handled the financial requirements for the Pittsburgh Works, such as payroll, which in the early days involved cash payments. On payday the Brinks truck would come in loaded with cash. What the people in the cashier's office had to do was take the payroll—and you're talking about millions of dollars—and break it down into different denominations so they would have the right amounts to fill all of the pay envelopes.

J&L switched over to checks in the late 1960s. The hourly people were accustomed to receiving cash, so it took a lot of doing to get the union to accept checks. Workers liked having cash in hand; they didn't have to go to the bank, and the cash was in their pockets ready to spend.

During the early 1970s Bill Roesch had just taken over as president of J&L. We had come off a loss year of about $20-some million, so we started cost-trimming programs. The loss grabbed everyone's attention. Under Roesch, there was very strict control and regimentation, monthly meetings in the board room downtown, and specific commitments made at each meeting to reduce costs, which in most cases meant reduced work force.

The loss was a combination of every-thing—supply and demand, the marketplace, and foreign competition. The work force was certainly part of the problem, because a good chunk of expense is labor costs.

In steel another big cost is investment. When you make a decision in steel, you've made one that's going to last for years. Every project that involved capital spending got a complete review of the expected economics, the investment, and the return on investment. The review was usually a joint function of industrial engineering and financial people. Our objective was to improve profitability with the investment we already had. This meant improving productivity, and that meant getting more production with fewer man hours.

What made the mill interesting was the vitality of the operation and the unexpected problems that arose. It seemed like there was always a challenging problem. Constant-ly. A furnace breaks down, a rolling mill breaks down, and you have to make some fast accounting and finance decisions, such as capturing the cost of the event. If insurance became involved, good documentation of the costs was vital. If the matter grew large enough, special funds might need to be requested.

In some businesses you could say, ''Let's have a committee meeting tomorrow and decide what to do.'' But steel people moved quickly. If you had iron breaking out of a blast furnace, you acted right then, not only because of the safety factor, but because it impacted on the departments in front and back of it: the coke-making department would have coke piling up, while the steel works waited on iron.

The steel business is big business. Each department in a steel plant is probably larger than most companies. For example, if you took the Pittsburgh Steel Works department, where they made about 150,000 tons of steel a month, and assumed a cost of $200 a ton, you're talking about a $30-million-a-month business. That's just to make the steel before it went through further processes.

Steel people never give up. The history of the business is that it runs in cycles, and everyone who works in the steel business knows and expects that. In the past the steel industry has always been able to bounce back strongly. For that reason, there was always hope that things would continue as usual. From a practical standpoint, when you saw the equipment getting older and older, foreign competition becoming stronger and stronger, sales prices being beaten down, and costs rising, it was apparent that some difficult adjustments had to be made.

To anyone born and raised in Pittsburgh, steel was Pittsburgh, and J&L was one of the top companies. Most native Pittsburghers thought it would be here forever.

Today, big mills have taken their licks. It is strange to see those flats along the river where there used to be a very active and teeming steel operation. An old J&L sign seen here or there is the only reminder of what used to be. Except for coke making and a galvanizing line that has been sold but is still operating, the Pittsburgh Works is gone. Except for a tin mill and a structural mill that has been sold but is still operating, the Aliquippa Works is gone. But we're over the tough part of the transition. As former Pittsburgh Works manager John Glasgow would say, ''The city is coming out of the mud.''

Ray Fleming

BELLS ARE RINGING, 1982

88

SNOW, TREE, 1983

FROM TRAIN SHED, SNOW, 1980

VIEW EAST, TOP OF POWER HOUSE, 1980

DICK FERRARI, FORMER INDUSTRIAL ENGINEER

Working at J&L taught me how to deal with confrontation. I had some good teachers! Most of the management and department superintendents were ''old style'' —tough and rough. Union leaders were the same; in fact, cut from the same cloth.

I had been on the job for about a month and a half when I was assigned to evaluate a request for new equipment in the strip mill. As I was walking into the site, this guy came over, looked at me, and said, ''Hey fellow, what are you doing here?''

''I'm here to review an equipment request.''

''You never called to ask my permission,'' he said.

''Well, who the hell are you?'' I asked.

He turned out to be the strip mill superintendent, so I put my hand out and introduced myself. Still he insisted that I couldn't enter without his approval.

''Let me put it to you this way,'' I told him. ''I'm here to evaluate your request for a new piece of equipment. I'll be glad to leave right now, but if I do, you won't get your equipment. I'll leave that up to you.''

Then this guy, who had a real gravelly voice, said, ''Okay kid, you're going to be all right. Go ahead.''

That was the first of my many ''tests,'' but all the supervisors were like that. They tried to intimidate the younger workers; union leaders did the same thing.

Industrial engineers like me were the link in the confrontational relationship between union and management. If management said we could do certain things, the union said the opposite. Industrial engineers were caught in the middle, even though we were officially management employees.

Among other things, our job required us to time operations to determine the efficiency level of equipment and personnel. From our observations, we could determine the correct amount of delay-time between activities and create accurate job descriptions. Incentives, job descriptions, past-practice provisions—all related to industrial engineering.

Generally, the workers we observed did not resent our being there. We always gave them advance notice of our arrival. We might run into trouble if they purposefully slowed down while we were studying a piece of equipment for performance capabilities. Sometimes their ploy worked, but most of the time we caught on.

One time when I was conducting a study at Eliza, a guy invited me to lunch. He took me to a little place on site called ''The Greek's,'' where a guy from outside the plant came in and made lunch every day. The menu consisted of fish, french fries, kielbasa, hamburgers, and hot dogs. Everything was cooked in the same pan of grease, and when it came out, it all tasted the same. But the food was great!

I stayed at the Pittsburgh Works until 1972, when I was reassigned to operations at the Cleveland Plant. Five years later I came back to Pittsburgh as superintendent of the strip mill. In the time I'd been gone, a change was evident. There was a difference in everyone's roles and interactions. Work forces had been reduced dramatically. Everyone knew the future of the plant was in jeopardy, so efforts had been made to reduce cost and increase efficiency. I was only there for a year the second time around, but I believe the attitude of labor, union, and management had improved. There was more communication on the state of business. Pluses and minuses were weighed, risks examined, and tasks outlined for the future.

In 1981 the hot strip mill rolled their last coil. I was working at Aliquippa at the time, but I made arrangements to go to Pittsburgh to see it. There must have been 100 people from J&L, not including the crew, that came to watch. Around 1 p.m., word came around that the last slab was going to drop out of the furnace.

I was walking along, talking with some of the mill operators as that last slab moved through the roughing and finishing stands to the coilers. The operator at the furnace drop-out area left his pulpit and walked along with us. The group sort of grew as the slab moved through its stages. We all held our breath, because the steel had to be rolled to a certain thickness in order to make medallions in honor of the event. We were worried that the steel would cobble, not roll properly. But when it got to the coilers, the operators got a good coil, and everybody went around shaking hands and congratulating each other. That last day was heart tugging, because there were people who had worked at the Pittsburgh Works for 30 or 40 years, and you could see the tears in their eyes. But they still had pride. Even on the last day they tried to keep productivity up and roll good steel. It gave you a good feeling to know such hard-working people.

The mementos they made that day said,
"Last coil on the 96-inch hot strip mill."
I've got one. That's something I'll always
hold on to.

Dick Ferrari

BLAST FURNACE, SNOW, 1980

97

POLITICAL BANNER STRING, 1982

VIEW SOUTH, CARS, 1982

FALLEN FURNACE, 1980

100

No one remembers precisely how the Eliza site was named. Some claim the choice was arbitrary, perhaps reflecting the word's Hebrew meaning, ''Consecrated to God.'' Others believe Eliza was a former mill manager's wife or daughter, a traditional source for blast furnace names. A few suggest that Eliza derives from the name of the geographical region where the furnaces were built, maybe a diminutive of Elizabeth, a nearby town named for the daughter and wife of Fort Pitt commander Colonel Aeneas Mackay and Revolutionary War officer Colonel Stephen Bayard, respectively. Whatever the mysterious origins of her name, Eliza would lead the burgeoning steel industry of nineteenth-century Pittsburgh to great heights.

Built in 1859 by Laughlin and Company along the north bank of the ''Steel City's'' Monongahela River, the two Eliza blast furnaces, the first of their kind in the region, produced a combined daily output of more than 150 tons of pig iron. Across the river stood Jones and Lauth (later Jones and Laughlins)

a steel company six years Eliza's senior. For decades the two mills worked in tandem; hot metal produced at Eliza was carried by ferry, later by railroad car, to the Southside plant for further refining. In 1900 the independent companies consolidated as Jones & Laughlins, Ltd., forerunner of the Jones & Laughlin Steel Corporation.

Over time additional blast furnaces were added to the Eliza site, older ones rebuilt, so that by 1916 a full complement of six furnaces existed. Their combined potential—more than 1.7 million tons of hot metal yearly—contributed to J&L's rapid growth during the early 1900s.

A blast furnace, so called because of the ''blast'' of hot air present in it during iron making, typically needs 3,300 pounds of iron ore, 1,200 pounds of coke, and 500 pounds of limestone to produce 2,000 pounds of iron. These raw ingredients, fed into the top of a tall, brick-lined metal chamber and then heated to more than 3,000° F., melt and react as they work their way to the bottom. Once fired, a furnace produces consecutive batches of iron until its heat-resistant bricks need replacing. One Eliza furnace made close to 3 million tons of hot metal before being ''put out of blast'' for repair.

As the American steel industry declined, so too did Eliza's output. One by one, her furnaces shut down, until by 1975 only the largest of the six remained in operation. Its last blast occurred at 10:45 a.m. on June 22, 1979, presaging doom for J&L's remaining Pittsburgh departments, who relied on hot metal from the Eliza furnace to continue operating.

When demolition of the Eliza furnaces ended in June 1983, a prominent landmark vanished from the Pittsburgh skyline. The only fully integrated, diversified steel mill in the city proper, J&L's Pittsburgh Works once employed close to 8,500 people, and in the 1960s the corporation itself ranked fourth in size among the nation's steel manufacturers.

Today, nothing of Eliza, except memories, and photographs, remains.

Many people helped make ELIZA possible. Although I can't list them all, I do want to gratefully acknowledge the following:

Donald Miller, whose elegy for Eliza interrupted my coffee on a crisp October morning nearly a decade ago. I took my camera to Eliza that very day and ended up leaving four years later.

David Bergholz, for his counsel and his careful review of the work in progress.

The Pittsburgh Foundation, The Pennsylvania Council on the Arts, and Bill Lafe, for providing grants and fellowships that assured this project would continue.

Jack Lane, for his unending enthusiasm for the work and his kindness and generosity in writing the introduction to ELIZA.

Seth Dickerman, who quietly shared his passion for making superb prints and, in doing so, revealed images that otherwise might never have been seen.

All those who knew Eliza best, those who worked there, and especially those who shared with us recollections of that time and place.

Ross, Kathleen, and Marilyn, for keeping a great faith, and Joan, for enduring with love and grace.

Mark Perrott

The editor wishes to extend a special thanks to the nine people who graciously shared their memories of Eliza here, so that others might remember too.